Topical Conference: Sustainable (Lower Carbon-Intensive) Energy Solutions: The Art of Scale-up and/or Scale-out

Held at the 2024 AIChE Annual Meeting

San Diego, California, USA
27-31 October 2024

ISBN: 979-8-3313-1669-3

Printed from e-media with permission by:

Curran Associates, Inc.
57 Morehouse Lane
Red Hook, NY 12571

Some format issues inherent in the e-media version may also appear in this print version.

Printed with permission by Curran Associates, Inc. (2025)

For permission requests, please contact AIChE
at the address below.

AIChE
120 Wall Street, FL 23
New York, NY 10005-4020

Phone: (800) 242-4363
Fax: (203) 775-5177

www.aiche.org

Additional copies of this publication are available from:

Curran Associates, Inc.
57 Morehouse Lane
Red Hook, NY 12571 USA
Phone: 845-758-0400
Fax: 845-758-2633
Email: curran@proceedings.com
Web: www.proceedings.com

TABLE OF CONTENTS

THERMOCHEMICAL PATHWAYS

103a The Role of Fusion in Highly Decarbonized Electricity Systems

Robert Armstrong

The urgent need to mitigate climate change requires drastic reduction in greenhouse emissions across the global economy. Emissions from the energy sector accounted for just under three-fourths of global greenhouse gas emissions in 2016[1], and this sector has thus been a major, early focus in reaching net-zero emissions goals. Because of the widespread availability of solar and wind resources globally and the dramatic decrease in costs of solar and wind conversion technologies over the past two decades, the electricity sector has been a major focus of decarbonization in the energy sector. These variable renewable energy (VRE) resources have grown rapidly to account for as much as 30% of generation capacity in California[2]. Because supply and demand must always be matched in the electricity sector, solar and wind must be complemented with storage technologies in order to get to high degrees of decarbonization. However, at emissions limits of 5-10 g-CO_2/kWh and below in the electricity system, VRE/storage options become very costly; and low-carbon, firm power options become important complements.

This presentation looks at the role fusion power plants could play in highly decarbonized electricity systems, given their zero-carbon, firm power attributes. Fusion power technologies are still under development, and there are not yet any commercial fusion power plants. Hence, we explore the potential role of fusion by investigating its integration into the electricity system in 2050 and beyond for a range of future fusion power plant costs. We look at fusion's value for different regions around the world as well as within the U.S. and conclude that fusion can play a major role both in the U.S. and globally. We also look at factors that the cost of fusion will be sensitive to in 2050 and beyond.

----

1. OurWorldinData.org. Sector by sector: where do global greenhouse gas emissions come from? - Our World in Data.
2. Data Show Clean Power Increasing, Fossil Fuel Decreasing in California. https://www.energy.ca.gov/news/2023-08/data-show-clean-power-increasing-fossil-fuel-decreasing-california#:~:text=The%20latest%20data%20show%20that%20in%202021%20more,an%20increase%20of%202.7%20percent%20compared%20to%202020.

Fusion

442a Challenges for the Design of Membrane Electrode Reactors for Use in Water Splitting and Electrochemical Reduction of CO_2

Alexis T. Bell

The rapidly decreasing price of renewable electricity has stimulated the development of electrochemical processes for producing chemicals and fuels. Two of the most highly explored areas are the splitting of water to produce H_2 and the reduction of CO_2 to produce ethene. Recent studies have demonstrated that to achieve current densities of > 0.1 A/cm^2 it is best to use membrane electrode assemblies (MEAs). An MEA consists of a polymeric cation- or anion-conducting membrane coated on one side with a cathode catalyst (for the generation of H_2 or the reduction of CO_2) and on the other with an anode catalyst (for the oxidation of water to O_2). An important measure of MEA performance is the kilowatt hours required to produce a kilogram of product. This metric is a strong function of the materials comprising the MEA and the conditions under which it is operated. This talk will illustrate how the design and operation of MEAs can be understood based on continuum models developed to simulate MEA performance and how such models can be used to guide their design to minimize the energy required to produce a kilogram of product.

103c Valuing and Deploying Industrial Energy Flexibility at Scale

Meagan S. Mauter

Significant variability in time of use electricity rates and high electricity demand charges are incentivizing industrial load flexibility and dispatchable generation. We propose a new set of metrics for valuing industrial load flexibility that enables direct comparison with other distributed energy storage technologies (e.g., batteries). Applying these metrics allows grid planners to assess the value of industrial flexibility to the grid relative to existing energy storage solutions and industrial operators to assess the financial benefits of energy flexibility for their facility operation. We apply these metrics to assess the benefits of load flexibility and dispatchable biomethane production with onsite electricity generation at water and wastewater treatment facilities. The framework reveals strengths and limitations of water system flexibility relative to other grid scale energy storage solutions, high value opportunities for flexible load operation of water assets, and the critical role of electricity tariff structures and energy service markets in determining water sector participation in load flexibility. Ultimately, this unified framework for valuating water asset flexibility enables both electricity and water asset managers to prioritize investments based on levelized cost comparisons across their respective portfolios.

103d Replacement of Metal Current Collectors with Carbon Nanotube Fabrics

Sibani Biswal

The global lithium-ion battery (LIB) market is poised for significant growth. Conventional current collectors, mainly composed of aluminum and copper foils, have remained unchanged since the inception of the first commercial lithium-ion battery. However, besides being critical materials, these current collectors contribute substantial weight to the electrodes (15-30%) without adding to the battery's capacity. Moreover, their production generates significant mining waste and emissions, highlighting the urgent need for sustainable alternatives.

I will describe the application of carbon nanotube fabrics (CNTf) as a replacement for copper and aluminum current collectors. CNTf possess properties that make them suitable for this role. Their semi-porous structure minimizes through-thickness electrical resistance in battery electrodes, while their axial alignment provides sufficient conductivity to extract current from large devices as effectively as metallic analogues. CNTf has the potential to significantly reduce the environmental impact of LIB production while enhancing their performance and sustainability.

103i Advanced Lithium Extraction for Battery Sustainability

Martin Bazant

Lithium was recently classified as a critical mineral, whose production must surge to meet growing demand for batteries without incurring unsustainable costs, carbon emissions or environmental hazards. Today, lithium is mainly produced by hard-rock mining and processed for battery manufacturing overseas, but much more lithium is available in brine resources, which mostly remain untapped for lack of an economical and sustainable extraction method. The conventional approach of solar evaporation ponds is not feasible for most brines due to their low lithium concentrations (<500ppm) and high impurity levels (> 10:1 Mg/Ca:Li), as well as the large water and land requirements (~100km2). Emerging methods of Direct Lithium Extraction (DLE) by adsorption or ion exchange have a smaller footprint but also struggle with these compositions and can require significant water or acids and bases. Moreover, neither these methods, nor existing hydro- or pyro-metallurgical processes for battery recycling, can economically extract lithium from black-mass for a circular economy. Advanced Lithium Extraction (ALE) based on electrochemistry has the potential to overcome these challenges by replacing harsh chemicals with electrons as reactants to achieve unprecedented lithium selectivity, while dramatically reducing water and energy use and carbon emissions, for a wide range of brines and leachates from mining and battery recycling.

296b Electrified Pathways to Ammonia at Ambient Conditions

Karthish Manthiram

Chemical synthesis is responsible for significant emissions of carbon dioxide worldwide. These emissions arise not only due to the energy requirements of chemical synthesis, but since hydrocarbon feedstocks can be overoxidized or used as hydrogen sources. Using renewable electricity to drive chemical synthesis may provide a route to overcoming these challenges, enabling synthetic routes which operate at benign conditions and utilize sustainable inputs. We are developing an electrosynthetic toolkit in which distributed feedstocks, including carbon dioxide, dinitrogen, water, and renewable electricity, can be converted into diverse fuels, chemicals, and materials.

In this presentation, we will share recent advances made in our laboratory on nitrogen fixation to synthesize ammonia at ambient conditions. Specifically, our lab is pursuing a lithium-mediated approach to ammonia synthesis and understood the reaction network that controls the reductive chemistry. The coupling of kinetics and transport has a significant influence on the rates of ammonia synthesis, providing a means to improve selectivity of nitrogen fixation. We have developed non-aqueous gas-diffusion electrodes which lead to high rates of ammonia synthesis at ambient conditions, overcoming transport limitations that would otherwise exist. Understanding of the solid electrolyte interphase through cryogenic transmission electron microscopy has provided unprecedented insight into the atomic structure of the reactive interphase. These findings will be discussed in the context of a broader range of electrosynthetic transformations which could lead to local and on-demand production of critical chemicals and materials.

103e Industrial-Scale Sustainable Electrochemical Systems

Saket Sanjay Bhargava

Replacing incumbent fossil fuel-based thermochemical processes with electrochemical technologies powered by renewable electricity is an emerging strategy for decarbonization of the chemical industry through electrification. Several chemistries (e.g., water electrolysis, carbon dioxide reduction, nitrogen reduction) have been thoroughly studied with a focus on new catalyst development, developing scalable reactor designs, and identifying the optimal operating conditions. Recent studies have also focused on understanding the underlying catalytic mechanisms and performing process systems analysis to understand economic and lifecycle feasibility. Improving electrode durability and system performance stability over time is still an active area of research. However, a systems thinking approach to understand what it would take for electrochemical systems to replace thermochemical systems is still missing – process design considerations, scaling rules, and sustainability aspects are integral for the development and deployment of industrial scale sustainable electrochemical systems. This talk will highlight the important process considerations, learnings that can be leveraged from allied fields (e.g., chlor-alkali process), and the key open questions and pressing research needs. Moreover, research on electrochemical systems has been mostly led by academia and national labs but the importance and urgency of academia-national labs-industry collaborations needed for technology maturation will also be highlighted.

296a Systems Perspective on the Role and Design Space of Emerging Hydrogen Technologies for Energy System Decarbonization

Dharik Mallapragada

As the emphasis of regional and national decarbonization efforts shifts from the power sector to the broader economy, there is growing interest in hydrogen (H_2) and enabling technologies for decarbonizing various end-uses where direct electrification remains challenged. The versatility of H_2, spanning production, storage, transportation and end-uses underscores its growing appeal. However, the many possible H_2 supply chains and the resulting cross-sectoral interactions complicates the assessment of H2 infrastructure and underlying technologies. In this presentation, we will highlight the use of advanced modeling frameworks to study cost-effective design and operation of H_2 technologies and supply chains for energy system decarbonization. The presentation will discuss the techno-economic outlook, emissions impacts and scalability of commercially available low-carbon hydrogen production pathways. Here, the importance of operational flexibility for electricity-intensive processes like water electrolysis to access low-carbon and low-cost electricity from an increasingly renewables dominant power system will be highlighted. At the same time, the design and scale-up implications of alternative electrolysis technologies will be highlighted. In addition, we will showcase the use of multi-vector energy system models to quantify the scale of H_2 technology deployment as well as the value and design space of emerging H_2 technologies.

296c Assessment of Low-Carbon Technologies in Energy Transition

Milind Deo

A number of new approaches are emerging as the world transitions to net-zero carbon. Use of hydrogen as universal fuel, carbon capture from concentrated point sources, and carbon dioxide removal (CDR) from the atmosphere are some of the possible technologies being implemented at scale. In this presentation, we assess the overall effectiveness of these technologies in achieving carbon neutrality. We also consider energy and water requirements of selected processes.

Momentum is building for a hydrogen-based economy. Large-scale use of hydrogen for a variety of industrial processes is common. In the current resurgence of interest in hydrogen, new methods of production, storage and end-use are being considered. Hydrogen is viewed as the fuel of choice to store excess renewable energy and use when dictated by demand. Energy requirements and efficiencies for common hydrogen production processes will be presented.

Renewable energy usage is also considered for capturing carbon dioxide from the atmosphere or from concentrated point sources, and for decarbonizing industrial processes. Energy requirements for established and emerging carbon capture and storage processes will be catalogued. Water usage over the life cycle of selected processes will also be presented. A combined assessment of energy and water usage along with carbon avoided will help highlight the net-zero carbon pathways.

296d High-Efficiency Perovskite Photovoltaics for Green Hydrogen at $1/Kg

Aditya Mohite

This presentation will cover three main areas of clean energy generation and conversion to value-added products. First, it will delve into efforts concerning the creation of high-efficiency, solution-processed perovskite solar cells. Here the primary focus has been on achieving stability that meets commercial standards, a significant hurdle thus far impeding widespread commercialization. The second portion of the presentation will include initiatives aimed at utilizing these solar cells for producing green hydrogen at a cost ranging from $1 to $2 per kilogram, accompanied by a thorough techno-economic evaluation. Lastly, the talk will introduce a new ongoing project, which entails the conversion of carbon dioxide and methane to value-added products through the application of innovative plasma chemistry techniques at scale.

296e Electrochemical Approach for Hydrogen Storage in LOHC

Lei Wang

Green-hydrogen is a promising candidate to address the pressing need of sustainable fuels. However, the intrinsically low mass-density (0.089 g/L) at ambient conditions makes its storage/transportation extremely challenging. Nowadays, hydrogen is mostly stored either as high-pressure-gas in cylinders/canisters or liquid in cryogenic tanks, both of which incur high containment costs, safety concerns, and boil-off loss.

Hydrogen storage in liquid-organic-hydrogen-carriers (LOHC) exhibits numerous advantages over the conventional compression/liquefaction technologies. Particularly, it can leverage the existing infrastructures for fossil fuels which are established in regards of both safety and cost. Besides, there are a handful of LOHC candidates that can fulfil the practical target of hydrogen-storage capacity, ≥6 wt% of entire systems including containers and valves. There are two steps for storing hydrogen in LOHC: an exothermic hydrogenation step followed by an endothermic dehydrogenation step. Recently, LOHC systems with significantly reduced dehydrogenation temperature of ~200 °C have been demonstrated, providing opportunities for the practical implementation of LOHC-based hydrogen storage. In contrast, the hydrogenation step, which also require relatively high temperature, but more importantly high H_2 pressure, remains a challenge due to the significant while inevitable entropy loss (gas to liquid). Here, we propose to tackle this challenge by direct hydrogenation of LOHC under ambient conditions electrochemically without phase transformation, using H_2O as the source of proton and electron, eliminating the safety concern posed by H_2-gas. Using a membrane electrode assembly-based reactor, and toluene as an example, we demonstrate the feasibility of this electrochemical approach for H_2 storage. Specifically, we achieve encouragingly high current efficiency (>95%) for the desired process of Toluene-Methylcyclohexane conversion, at industrial relevant current density. Furthermore, based on the preliminary data collected, we conducted comprehensive Life-cycle-assessment and Technoeconomic analysis to identify the technological gaps for the implementation of this technology.

296f Hydrogen Supply Scale-up Solutions for Industrial Decarbonization

Gaurab Samanta

Power generating processes are some of the most common in industry today. Impactful scope 1 emissions reduction can be achieved through their decarbonization. Hydrogen is a powerful scalable enabler of decarbonization of industrial power generating processes. For decarbonization goals to be successful, it is critical to understand appropriate supply solutions to optimally unlock the scalability potential of hydrogen. We will discuss new innovations to enable the decarbonization of boilers and power generators such as gas turbines. We also will discuss supply chain optimization to maximize the benefit of the traditional utilities listed above plus distributed energy/micro-grid processes such as internal combustion engines and fuel cells. In each scenario, we will consider operational requirements, opportunities for incremental reduction of emissions enabled through hydrogen, supply mode applicability, consistency, and redundancy for sustained reliable performance.

296g Multi-Tier Analysis of Low-Emission Hydrogen Production

Zitao Wu, Haibo Zhai, Eugene Holubnyak, Selena Gerace, Curtis Biggs

The U.S. government is prioritizing the transition to clean energy due to concerns about climate change and energy security. Meanwhile, clean hydrogen is gaining attention for its high energy density and low lifecycle greenhouse gas (GHG) emissions. In 2022, President Biden signed the Inflation Reduction Act (IRA), providing funding and incentives to foster a clean energy economy, including incentives for low-emission hydrogen production (such as Section 45Q and Section 45V). In the meantime, the U.S. leads global natural gas (NG) production, but concerns arise due to associated GHG emissions. Steam methane reforming (SMR) with carbon capture and storage (CCS) is one of the low-emission hydrogen production methods, and emerges as a potential economic opportunity to utilize abundant resources and promote environmental responsibility in fossil-rich regions. Wyoming is one such state that has the potential of becoming a key player because of its abundant fossil energy resources, CO_2 storage resources, and natural gas and CO_2 transmission infrastructure. In addition, the state has made significant efforts in decarbonizing the state's energy sector with policy and regulations, such as HB200 Low-Carbon Energy Standards. These resources, infrastructures, and regulations establish a solid basis for the development of an advanced hydrogen economy in the state of Wyoming.

This study reveals the interconnections among the techno-economics and resource impacts of low-emission hydrogen production under the IRA incentive policies in a fossil-rich state. The objective is to assess the feasibility of low-emission hydrogen production within the state. The research findings are intended to provide valuable insights into energy management decisions and policies as the energy sector transitions toward low-emissions, particularly in fossil-rich regions. To achieve the research goals, this study is conducted from various aspects at both plant-level and cluster/state-level as described below.

> *Plant-level techno-economic analysis.* This study evaluates the technical and economic dimensions of a gas-based commercial hydrogen production plant, employing SMR with CCS, in line with the National Energy Technology Laboratory (NETL)'s baseline study. Performance metrics such as consumption intensities of water, natural gas, and electricity, as well as CO_2 capture rate, are assessed for the hydrogen production plant. Moreover, an engineering-economic model is developed to estimate the levelized cost of hydrogen (LCOH) production, including capital costs and operating & maintenance (O&M) costs, incorporating the performance metrics. The costs are adjusted to be specific to a region. The capital costs are adjusted to local values, using a regional adjustment factor reflecting regional variations in ambient conditions, labor, and material markets. O&M costs incorporate local resource and utility prices, as well as labor expenses. Furthermore, uncertainty analysis is conducted, encompassing the uncertainty in plant design and financial and economic parameters, using the Latin Hypercube Sampling (LHS) method to estimate LCOH in a probabilistic form.
>
> *IRA impact on hydrogen production cost.* The hydrogen plant can either claim the 45V tax credit or the 45Q tax credit and is presumed to fulfill registered apprenticeship requirements. Thus, the bonus rate of IRA 45Q tax credit is \$85 per metric ton CO_2 captured and stored in saline reservoirs and \$60 per metric ton CO_2 captured and utilized by enhanced oil recovery (EOR) over a period of 12 years. At the same time, the bonus rates of IRA 45V tax credits range from \$0.6 to \$3.0 per kg hydrogen produced over a period of 10 years according to the

tiers of life cycle GHG emissions (limit of 4.0 kg CO_2-eq/kg H_2). According to the annual amount of captured CO_2 and hydrogen production capacity, the total tax credits are summarized over the tax period, then levelized to the overall hydrogen production throughout the plant book lifetime.

Lifecycle GHG emissions. This study develops a cradle-to-grave LCA model to estimate GHG emissions of hydrogen production across the fuel cycle to determine the eligibility for and rate of the 45V tax credit. The Intergovernmental Panel on Climate Change (IPCC) Fifth Assessment Report (AR5) global warming potentials with carbon-cycle feedback for a 100-year time frame are used to convert non-CO_2 emissions into CO_2 equivalent (CO_2-eq) emissions. The LCA system boundary encompasses upstream NG and electricity supply, the hydrogen production plant, and downstream CO_2 sequestration. Both fossil and renewable NG (FNG and RNG) supplies are analyzed in this study, with emission factors of regional FNG supplies obtained from Littlefield et al. (2022). GHG emissions of RNG produced from feedstocks, including animal manure, landfill gas, and wastewater sludge, are estimated using Argonne National Laboratory's (ANL) GREET model. The emissions avoided from the current management methods of the feedstocks are considered as the credit of life cycle emissions of RNG supply. Additionally, GHG emissions of local power grid and alternative power sources (e.g., solar, wind, nuclear) are also modeled using the GREET model based on the local electricity generation profile. GHG emissions of CO_2 storage in saline reservoirs are estimated based on the study of Skone et al. (2013). This study estimates major components with emissions larger than 0.01 kg CO_2/tonne CO_2 sequestered and incorporating region-specific emission factors. In contrast, the emissions of EOR activities that purchase CO_2 from the hydrogen plant are not accounted for in the life cycle emissions. Furthermore, uncertainty analysis is conducted, encompassing the uncertainty in the NG and electricity supply as well as CO_2 storage using the LHS method to estimate life cycle GHG emissions in a probabilistic form.

State-level resource impact analysis. The development of a low-carbon hydrogen cluster at the state level requires various resources. The annual production capacity of FNG is obtained from the U.S. Energy Information Administration, while the potential for RNG production can be estimated based on feedstock (animal manure production, landfill waste in place, and wastewater flow rate) availability and methane conversion efficiencies. State-level hydrogen production potential can be estimated as a function of the annual production capacity or potential of both FNG and RNG according to the plant-level NG consumption rate. Additional resource requirements, including water, electricity, and CO_2 storage, are then estimated using plant-level performance metrics relative to the annual hydrogen production capacity. ArcGIS Pro is utilized to spatially analyze suitable locations for hydrogen plants based on the co-availability of multiple resources and infrastructures, such as freshwater availability, saline storage resources, oil and gas fields, and natural gas and CO_2 pipelines.

Preliminary Results. The performance, economics, and life cycle emissions of a low-carbon hydrogen plant in the context of IRA's tax credits are assessed at a plant level. The resource use and infrastructures to support cluster-level development is further explored at the state level. The LCOH ranges from \$1.6 to \$2.0 per kg H_2 for SMR with CCS utilizing FNG and electricity from the power grid. In comparison, the hydrogen produced with CO_2 used for EOR can reduce the LCOH to \$1.1–1.6/kg H_2 through the revenue generated from CO_2 sales. The 45Q tax credit can reduce about 19% of the LCOH. Compared to 45Q tax credit, the 45V tax credit has less of an effect on reducing the production cost. This is because the probability of life cycle emissions lower than 4 kg CO_2-eq/kg H_2 is less than 6% and 11% for the hydrogen plant with CCS and EOR, respectively. NG and electricity supply are the largest two emission components. Using alternative electricity sources (e.g., renewable energy or nuclear power) can reduce life cycle emissions. However, only the lowest tier of 45V tax credit is applicable, still less than that of the 45Q tax credit. RNG (animal manure and wastewater sludge) instead of FNG can achieve negative life cycle emissions, thus helping the hydrogen plant obtain the maximum level of 45V tax credits. However, the price of RNG is much higher than FNG and the low production potential makes RNG currently uncompetitive in Wyoming. With the expansion of production capacity and technology development, LCOH can further decrease with learning-by-doing. In Wyoming, 20% of the marketed production of NG can support a low-carbon hydrogen production capacity of 1.24 million metric tons per year, necessitating less than 0.4% of the state's total water withdrawal and less than 10% of interstate electricity export. In addition, the hydrogen cluster can leverage the existing and planned CO_2 pipelines and abundant existing natural gas infrastructures, as well as CO_2 storage resources. This study demonstrates that, with the support of IRA policies and state resources, the development of low-emission hydrogen economy can be sustained in Wyoming. Supportive policies at the state-level can further increase the competitiveness of low-carbon hydrogen when compared to fossil energy. Therefore, the application of low-emission hydrogen in other sectors such as transportation, utility, and industry can foster a low-emission economy in Wyoming and surrounding areas. The analysis framework is applicable to other fossil-rich states and offers insights into decision-making and policy designs for transitioning toward a low-emission energy economy.

442b Fast Track Technology De-Risking, Scale-up and Commercialization Via Hybrid Simulation – Laboratory Virtual Piloting and Demonstration

Joseph Powell

This paper will examine the combination of fit-for-purpose bench scale laboratory experiments combined with digital simulation to enable direct scaling from bench to commercial, limiting the need for continuous piloting and demonstration scale-up for many systems. The methodology can be used for early identification of preferred process concepts on the basis of ease and speed of scaling, with many examples of unit operation prototypes for reaction and separation. Examples of where fast track development was successfully deployed will be provided, along with examples where process and technology required traditional slower pathways through full piloting and separate demonstration.

442c Synthetic Biology for Energy Applications

Ariel Furst

Significant effort has been devoted to developing technologies that effectively mimic biological processes, but these methods often fail to replicate the efficiency and selectivity of native systems. We have found that, by combining electrochemistry with the inherent activity of microbes, we can improve upon conventional technologies for sustainable energy production and environmental remediation. Using synthetic biology, we can engineer microbes to both degrade and detect harmful environmental pollutants, synthesize catalysts for greenhouse gas valorization, and perform green chemical conversions. Through these technologies, we have consistently found that the combination of synthetic biology and chemistry affords advantages beyond the capabilities of either technology alone.

442d Prototyping and Scale up of Electrochemical Reactors for CO_2 Reduction

Meenesh R Singh

This talk delves into the development and optimization of model-driven approaches for Gen-1 (50 cm^2) and Gen-2 (500 cm^2) electrolyzer designs aimed at achieving targeted performance levels in CO_2 conversion. The scale-up methodology for the zero-gap electrolyzer focuses on maintaining kinetic and geometric similarity between small-scale and scaled-up systems. As electrochemical reactors are primarily planar, scale-up significantly impacts mass transfer, leading to surface heterogeneities and product distribution variations. To mitigate these effects, cells are restricted to sizes below 2000 cm^2, with desired production rates achieved through cell stacking. The scale-up philosophy emphasizes maintaining similarity between individual cells within a stack to ensure uniform voltage distribution. This necessitates the use of modeling and simulation tools for evaluating kinetic and geometric similarity. The first task is to evaluate potential and concentration distributions in small-scale electrolyzers for assessing reaction performance, followed by assessment of current distribution, pH gradient, and potential along the electrode surface in Gen-1 and Gen-2 electrolyzers, respectively. Through these efforts, we aim to optimize electrode designs and operating conditions for maximum CO_2 utilization rates with minimal power consumption. Additionally, the talk highlights strategies for overcoming mass and charge transport challenges in larger-scale electrolyzers while ensuring high activity and selectivity of catalysts. Furthermore, the development of a Multiphysics model-based diagnostic tool enables real-time monitoring and assessment of electrolyzer performance, facilitating efficient and reliable CO_2 conversion processes.

442e Challenges in Practical Electrochemical CO2 Conversion That Are Beyond Catalysts

Haotian Wang

Electrochemical carbon dioxide conversion to valuable chemicals and fuels has been widely recognized as a promising approach for chemical manufacturing decarbonization. While extensive research and development has been focused on catalytic materials design, there are challenges beyond the level of catalysts for the successful scaling up and deployment of this technology. In this talk, I will use some examples to discuss these challenges and how we can address them.

442f Scalable Materials Solutions for the Deployment of Adsorption-Based Direct Air Capture Technologies

Praveen Bollini

The anthropogenic removal of CO_2 from ambient air using direct air capture (DAC) technology is receiving increasing attention recently in light of the need for negative emissions technologies that reverse current trends in atmospheric CO_2 concentrations. Capture materials considered thus far for the most part comprise organic components in addition to inorganic ones, with the former resulting in inherent limitations in hydrothermal and oxidative stability. I discuss here how materials that are purely inorganic in composition can capture CO_2 reversibly under ambient conditions in the presence of humidity, thereby significantly simplifying process complexity and cost. Unlike most inorganic sorbents that are either intolerant to water or require regeneration temperatures in excess of 300°C, mixed metal hydroxides can capture and release CO_2 with a less than 100°C swing in temperature. They are also exceedingly hydrothermally and oxidative stable, as reflected in cycling experiments with steam and oxygen. This talk will discuss how this novel class of DAC sorbents circumvents challenges relating to material precursor availability, cost, and stability. Potential contactor designs for scaling up DAC will also be discussed.

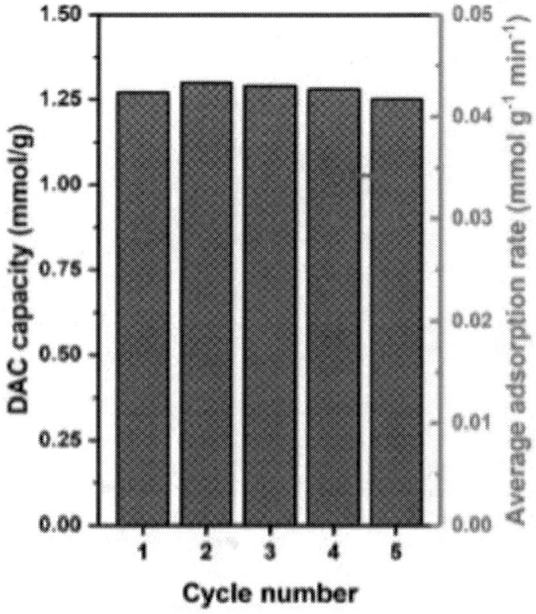

442g New Process Integration of CO_2 Capture and Utilization for the Ironmaking Process

Wei Wu, Po-Chih Kuo, Yin-Ping Wu

The blast furnace (BF) is not only a complicated metallurgical process, but it induces a large amount of CO_2 emissions due to the coke and coal as sources of reducing agents for iron ore reduction. A new carbon capture and utilization (CCU) approach is connected to the top gas of BF system, which aims to convert the captured CO_2 into the methanol fuel through the co-electrolysis of steam and CO_2 in a solid oxide electrolyzer cell (SOEC). The external power supply is based on an integration of the coke oven plant and the solid oxide fuel cell (SOFC). The proposed integrated ironmaking process is depicted in **Fig. 1**. The thermodynamics of chemical metallurgy for the BF model is extracted from FactSage® and the BF model validation is achieved by using Aspen Plus®. The heat exchanger network is involved to improve the heat recovery of the integration of BF and CCU. It is validated that the total CO_2 emissions of the BF with CCU is lower than the BF without CCU by 33% and the energy conversion efficiency of the CCU is close to 30.9%.

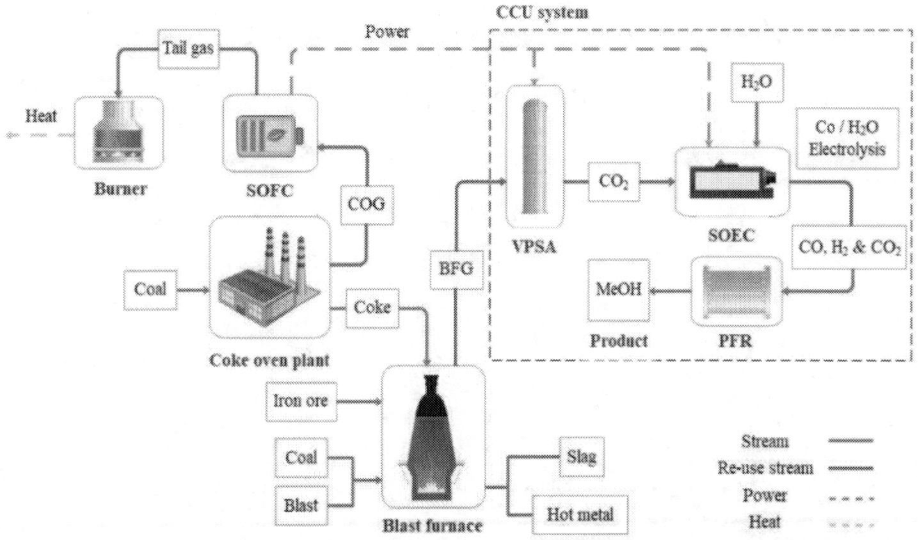

Fig. 1 Flowchart of the blast furnace (BF) integrated with the coke oven plant, SOFC, and CCU

103f Six Learnings from Product and Process Development Applied to Decarbonization Scale-up/Scale-out

Gayle Gibson

All chemical or material processes need energy. As we seek to decarbonize current processes and commercialize new ones, the scale-up/scale-out challenges can be daunting. As engineers, we see technical solutions. This presentation will pose 6 areas of learnings from recent startups as they seek to scale-up/scale-out. Some of these are from decades of product and process development and others are more unique to today's decarbonization efforts.

236a Chemical Looping Technology for Clean Energy Conversion Systems – Commercialization Prospect

Liang-Shih Fan

The concept of chemical looping reactions in redox operating schemes has been recognized as a valuable means in the synthesis of fuels and chemicals in industries. Fundamental research on chemical looping reactions has also been conducted for many decades. The well-recognized chemical looping reactions could be represented by the steam-iron process using coal from the late 1800s to early1900s and were applied at a pilot scale for synthetic natural gas production with the IGT HYGAS Process and for syngas production with the CO2 Acceptor Process in the 1960s and 1970s. There are presently, however, no chemical looping processes in commercial operation. Key technical factors that determine commercial viability of the technology lie in the sustainability of the reactivity and recyclability of the metal oxide oxygen carriers and the ability of configuring the reactor assembly for optimal product yields. The chemical looping system is developed on the circulating fluidized bed platform. The successful deployment of this technology requires thorough knowledge of two interconnected fields, i.e., metal oxide reaction engineering and particle science and technology. With now CO2 emission control of great concern and process conversion efficiency enhancement of great interest, activities on research and development of chemical looping technology have resurfaced. Specifically, chemical looping as a platform technology is a manifestation of the interplay among such key elements of metal oxide reaction engineering and particle science and technology as particle synthesis, reactivity and mechanical properties, flow stability and contact mechanics, gas-solid reaction engineering and particulates system engineering. This presentation will describe the fundamental and applied features of modern chemical looping technology in the context of the circulating fluidized bed platform that utilizes fossil, biomass and other feedstocks. It will discuss the reaction chemistry, ionic diffusion mechanisms, metal oxide synthesis and thermodynamics, reactor design, and system engineering along with energy conversion efficiency and economics of the chemical looping processes for, specially, partial and selective oxidation for hydrogen, syngas and chemicals production. The Ohio States University has developed a number of advanced chemical looping gasification and reforming processes which have been licensed to industries for commercialization. This presentation will highlight the readiness of this technology commercialization.

103h Distributed Direct Air Capture for Dense Urban Spaces

Mei Chee Tan

Existing efforts to reduce the carbon footprint of the energy grid alone is inadequate to meet net zero emissions targets by 2050. To achieve net zero goals, any released CO2 emissions must be balanced by the permanent removal of CO2. Subsequently, in the longer term, CO2 sources used in CO2-derived products will largely be from the atmosphere. The technology that actively captures of CO2 directly from the air is known as direct air capture (DAC). The mainstream implementation of DAC technology has been limited by: (1) high energy penalties, (2) process scaling risks, and (3) land-intensive requirements. Consequently, the aforementioned factors pose a substantial barrier for the deployment of DAC technologies, especially in dense, urban cities. Taking into account these constraints, our team has developed a material driven solution that would facilitate the rapid deployment of DAC in high rise buildings. Compared to building a centralized DAC facility with a singular function of removing CO2 from the air, the integration of DAC hubs within buildings would enable DAC to be implemented in a distributed, scale-out manner. Using the distributed direct air capture (DDAC) approach, ambient CO2 levels could be reduced at urban city states through the collective deployment of the DDAC technology. Besides potential air quality improvements, the elevated indoor CO2 levels presents a thermodynamic advantage for DAC implementation. Similarly, building energy efficiency could benefit from increased circulation of air within the heating, ventilation and air conditioning (HVAC) systems. The coupling of HVAC and DAC systems could allow for the recirculation of air within the building instead of repeatedly and inefficiently intaking fresh outdoor air.
Once captured within our proprietary sorbent, CO2 is released and the sorbent is regenerated at a centralized facility. The later use of CO2, either permanently stored (e.g., within concrete or geological formations) or industrial feedstock, will dictate the required product gas composition and pressure. The energy requirements of DDAC plants are thus associated with applicationspecific factors that are highly context dependent, calling for case-by-case life-cycle assessments. Informed by these assessments, a full specification required for the circular carbon ecosystem is established. For instance, the optimal and viable transport mode for CO2 and other practical logistic and cost challenges for the different applications much be evaluated.

236b Carbon Dioxide Sequestration Via Gas Hydrates: A Potential Pathway Towards Decarbonization

Praveen Linga

Climate change is known to be dominantly caused by the increased concentration of greenhouse gases in the atmosphere, in particular carbon dioxide (CO_2). The clathrate hydrate process has been demonstrated over the years as promising technology for innovative applications like natural gas storage, carbon dioxide capture, seawater desalination, cold energy storage etc. CO_2 hydrate, a solid compound made of molecular CO_2 enclathrated in crystalline lattices formed by water molecules, is an attractive option to capture and for long-term CO_2 sequestration. Methane hydrates in oceanic sediments have been stable for millions of years. As a natural analog, is it possible to store CO_2 in the form of hydrates in oceanic sediments forever? In this presentation, the state of the art on clathrate hydrate technology pertaining to carbon capture and storage will be discussed in detail. Furthermore, future research and development opportunities and pathways for commercialization will be discussed.

236c Scaling up Medium-Temperature Reverse Water-Gas Shift Catalysis

Matthew Kanan

Reverse water-gas shift (RWGS) provides a critical link between established water electrolysis and syngas-to-liquids technologies to enable integrated power-to-liquids (PtL) production. Current (pre)commercial RWGS technologies utilize Ni catalysts that are derivatives of steam reforming catalysts. These materials catalyze not just RWGS but also methane (CH_4) and coke formation. To suppress methane and avoid coking, conventional Ni RWGS catalysts must be operated at very high temperatures (typically 900-1100 °C) where RWGS is thermodynamically favored over these competing reactions. This temperature regime necessitates expensive materials of construction and complicates heat integration with the rest of a PtL process. We have developed RWGS catalysts that are composed only of alkali carbonates (e.g. K_2CO_3) dispersed in mesoporous supports such as Al_2O_3 ("dispersed carbonate RWGS catalysts). Because they have no transition metals, dispersed carbonate catalysts are extremely selective for RWGS, enabling operation in a medium-temperature regime without CH_4 or coke formation. In independent testing, high activity and essentially 100% selectivity for CO has been achieved at industrially relevant gas hourly space velocity (>10,000 h^{-1}) at ≥500 °C with no degradation of performance observed over multiple weeks. The ease with which dispersed carbonates can be synthesized enables ultra-low-cost manufacturing using existing supply chains and production facilities. This talk will describe the development, state-of-the-art, and scale-up of dispersed carbonate RWGS catalysis and their application in sustainable chemical and fuel synthesis.

236e Research in Low-Carbon Technologies and Alternative Feedstocks for Chemicals and Energy

Lili Zhang

Abstract

Carbon negative and energy efficient technologies that could reduce greenhouse gas (GHG) emissions are among the greatest challenges of the 21st century. Therefore, developing processes and technologies with low GHG emissions to meet growing energy demand and to produce chemical and materials in a sustainable manner is of great significance. Multifunctional catalysts with atomically dispersed active centers are highly desirable in advancing the development of sustainable processes and technologies with atom economic and environmentally friendly methods for the production of chemicals and materials. However, how to precisely construct active catalytic sites with precisely defined coordination environment having favorable structure at the atomic level to achieve targeted products remains challenging. What's more challenging is how to mass produce advanced catalyst materials with highly active centers. In this talk, I will first describe how we achieved multifunctional catalysts with good catalytic performance for CO2 conversion, clean energy production and waste valorization. Then I will talk about how use-inspired basic research is coupled with accelerated catalyst development platform to accelerate the development of the emerging technology, and shorten the time from lab to market.

236f Recent Advances and Prospects of Hydrate-Based Pre-Combustion Carbon Capture

Nagu Daraboina

Carbon capture, utilization, and storage (CCUS) technologies play a crucial role in delivering clean energy and combating climate change, particularly as electricity demand escalates alongside population growth and the rapid expansion of the electric vehicle market. A newly developed approach for precombustion carbon capture utilizing gas hydrates has drawn considerable attention because of its promising potential.

Hydrate-based precombustion carbon capture is a method that involves using gas hydrates, which are ice-like crystalline solids formed by water and gas molecules under specific pressure and temperature conditions, to capture carbon dioxide (CO_2) from fuel gases before combustion. The benefits of hydrate-based precombustion carbon capture are high capture efficiency, low energy consumption, scalability, the potential for CO_2 storage as hydrates, and compatibility with various fuel types. The major bottleneck that was impacting the scaling up of the hydrate-based precombustion carbon dioxide capture process is the slow kinetics of hydrate formation.

In this presentation, we present the overview of hydrate-based precombustion carbon dioxide capture from fuel gas mixture and our current research efforts to improve the kinetics by employing different reactor configurations. We also present the future directions to commercialize this technology.

236g Scale up Analysis of Inductively Heated Metamaterial Reactors

Chenghao Wan, Connor Cremers, Zhennan Ru, Calvin Lin, Kesha Tamakuwala, Dorothy Mantle, Ariana Hofelmann, Pinak Mohapatra, Matthew Kanan, Juan Rivas, Jonathan Fan

The electrification of heavy industry is critical in the broader effort to decarbonize chemical manufacturing, and the scaling of electrified thermochemical reactors is particularly crucial to meeting industrial demands. Methods such as resistive and microwave heating reactors face considerable scaling challenges in construction feasibility, heating volume, and efficiency, limiting their large-scale deployment for many applications. Induction heating, on the other hand, is an established concept in the metallurgy industry, and industrial induction furnaces are technologically mature and operate at megawatt power levels. For chemical reaction processes, induction heating is further distinguished by its potential for volumetric heating, safe operation, and ease of implementation.

In this work, we explore the potential for scaling up inductively heated reactor systems using a combination of analytical modeling, numerical simulations, and experiments. We will show that heating efficiency, defined as the conversion of external electricity to energy consumed by the chemical reaction process, increases towards values over 90% as the reactor scales up, and that heating efficiency is fundamentally limited by AC-DC electrical power conversion efficiency. This study builds on previous work where we demonstrated a high-efficiency induction-heating reactor termed a metamaterial reactor. Our prototype reactor has a 1.5-inch diameter and utilizes an open-cell SiSiC foam baffle that simultaneously serves as a heating susceptor and catalyst support. Proper tailoring of SiSiC foam's effective electrical conductivity and induction frequency led to reactor conditions supporting uniform radial temperature distributions and a heating efficiency 87%, and we showed that the reactor could be used to perform the reverse water-gas shift (RWGS) reaction across a temperature range of 500 – 650 °C with ideal plug flow behavior.

To analytically deduce efficiency scaling trends with metamaterial reactors, we develop a model that investigates power consumption within the reactor system, as decomposed by useful contributions (i.e., heating of the inlet gas stream and reaction enthalpy) and parasitic contributions (i.e., energy loss via thermal conduction, induction coil heating, and energy dissipation in the amplifier). In our model, we consider the scaling of each linear dimension of the reactor (i.e., diameter and height) by a factor , and we adjust gas flow rates to keep mass flow per reactor volume constant. Utilizing our experimental data and analytical expressions to evaluate each component, we establish that these factors scale differently with —gas heating, reaction enthalpy, and amplifier loss are proportional to ^{3}, coil loss is proportional to ^{2}, and thermal conduction loss is proportional to . As a result, when projecting these effects for larger reactors, our model anticipates an enhanced overall energy efficiency surpassing 90% when reactor radius approaching 0.25 m, and that the cap in total efficiency is set by the efficiency of the power amplifier.

Numerically, we construct a multiphysics model (implemented in COMSOL Multiphysics) to further investigate the temperature distribution and power consumption characteristics of our reactors. These models combine and couple electromagnetic, heat-transfer, and computational fluid dynamics simulations. Our simulation is first constructed and refined to fit our experimental results and analytical modeling. The model is then extended to scaled up reactor systems to investigate how the heating and temperature profiles evolve as the reactor scales up. The simulations reveal that our inductively heated reactors are capable of sustaining a uniform radial temperature profile and enhanced efficiency as it scales, due to its ability to support volumetric heating capabilities. We also analyze wall-heated reactors and find that they support radial temperature gradients that become increasingly non-uniform as the reactor is scaled up.

To complement these theoretical findings, we experimentally construct reactors with diameters of 1.5 inches and 3 inches and perform the RWGS reaction within each reactor. In this scale up effort, the effective conductivity (σ) of the SiSiC baffle is adjusted such that the penetration depth of the magnetic field into the reactor (skin depth, $\delta = \sqrt{(1/\pi\sigma f\mu_0)}$) scales with the reactor radius. Through an energy consumption analysis and the experimental characterization of temperature profiles within the reactors, we show that the heating efficiency of the reactors increase with reactor diameter in a manner that agrees well with analytic and numerical modeling.

AUTHOR INDEX

AIChE
120 Wall Street, FL 23
New York, NY 10005-4020

ISBN 979-8-3313-1669-3

CURRAN ASSOCIATES INC.
proceedings.com